ENDANGERED!

SEALS

John Woodward

Series Consultant: James G. Doherty
General Curator, The Bronx Zoo, New York

BENCHMARK BOOKS

MARSHALL CAVENDISH
NEW YORK

Benchmark Books
Marshall Cavendish Corporation
99 White Plains Road
Tarrytown, New York 10591-9001

Library of Congress Cataloging-in-Publication Data

Woodward, John, 1954-
　　Seals / by John Woodward.
　　　　p.　cm. — (Endangered!)
　　Includes index.
　　Summary: Describes the physical characteristics, habitat, and
behavior of seals and sea lions, and points out that there are still
many threats to their survival.
　　ISBN 0-7614-0292-6
　　1. Pinnipedia—Juvenile literature.　2. Seals (Animals)—Juvenile
literature.　3. Sea lions—Juvenile literature.　4. Endangered
species—Juvenile literature.　[1. Seals (Animals).　2. Sea lions.
3. Endangered species.]　I. Title.　II. Series.
QL737.P6W66　1997
599.74'5—dc20
　　　　　　　　　　　　　　　　　　　　　　　　　　96-7223
　　　　　　　　　　　　　　　　　　　　　　　　　　CIP
　　　　　　　　　　　　　　　　　　　　　　　　　　AC

Printed and bound in the United States

PICTURE CREDITS
*The publishers would like to thank the following picture libraries for supplying
the photographs used in this book:* Bruce Coleman Ltd FC, 4, 5, 6, 8, 9, 13, 14,
15, 16, 18, 19, 23, 24, 27, 29, BC; Frank Lane Picture Agency 1, 7, 11, 12, 20,
22, 25, 28.

Series created by Brown Packaging

Front cover: Harp seal pup.
Title page: Young southern elephant seal.
Back cover: Fur seal and elephant seal.

Contents

Introduction

If you were trying to describe a seal, you might say it looked like a dog with the body of a porpoise. You would not be far from the truth. Long ago, seals were animals that looked a bit like bears. They started hunting in the water instead of on land, and over millions of years they **adapted** to their new way of life. They developed flippers in place of feet, and their bodies became more streamlined.

However, beneath their fishy disguise seals are **mammals**, as are dogs, cats, and human beings, among many others. Because of this, they cannot breathe underwater and must come to the surface to gulp fresh air.

A Weddell seal relaxes on the Antarctic ice. Seals have a thick layer of fat under the skin that keeps them warm in the water and enables them to live in cold places.

Most seals hunt fish and squid, and they have sharp, spiky teeth for gripping their struggling, slippery **prey**. The Weddell seal is a fish eater, but it will also feed on small animals that live on the seafloor. One of the deepest-diving seals, the Weddell can go down as far as 2000 feet (600 m) in search of food.

Some seals living in the Antarctic feed on the shrimplike krill that swarm in the Southern Ocean. One **species**, called the crabeater seal, has sievelike teeth for straining krill out of seawater. The crabeater is hunted by the leopard seal – the only seal that regularly kills other seals for food.

The leopard seal is a large seal that can reach more than 9 feet (2.7 m) in length. Besides hunting other seals, it kills and eats penguins.

Introduction

There are three main groups of seals: the "true" seals, the eared seals, and the walrus. Today many of these animals are protected by law, but in the past several species were almost wiped out by hunters. Seals were slaughtered for their fur, which was used to make clothes, and for their fat, which was turned into oil. Although some hunting still takes place, seals now also face other threats, such as **pollution** and overfishing. Even so, only two species are in serious danger of becoming **extinct**. Others are rare, though, and some seal populations are getting smaller every year. Scientists are working hard to find out why, before it is too late.

In this book, we will look at each of the three groups of seals, beginning with the true seals.

A fur seal (left) meets an elephant seal in the Antarctic. A crowd of king penguins take no notice as the two seals defend their breeding areas on the beach.

True Seals

True seals are the most fishlike of the seals. A true seal's hind flippers face backward like a fish's tail and are used to drive the seal through the water. Its front flippers are fairly small and handlike, and the seal sometimes steers with these as it swims. Both sets of flippers are almost useless on land, though. They cannot bear the animal's weight, so when a true seal comes ashore it has to drag itself around on its belly. There are 18 species of true seals, including the Weddell, crabeater, and leopard seals. Here we will look at some true seals that are at risk: the harp and ringed seals, the monk seals, and the elephant seals.

Although they are slow and clumsy on land, true seals, like this Mediterranean monk seal, can swim and dive like fish.

Harp & Ringed Seals

One of the best known of the true seals is the harp seal.
This species lives throughout the Arctic from Hudson Bay
to the Laptev Sea off Siberia. An adult harp seal can grow
to about 5 feet 7 inches (1.7 m) long and is generally
silver-gray with a dark face and head. Harp seals also have
dark markings on their backs. The seals were named after
these markings, which reminded people of a harp. In fact,
especially in males, the pattern forms more of a U-shape.

Harp seals spend most of the year spread throughout
their **range**, but in spring they gather to breed. Unlike

*Harp seals
hunt fish under
the Arctic ice,
surfacing
through holes
and cracks
when they
need to
breathe.*

whales, which are also mammals, seals must come out of the water to give birth. Their clumsiness on land makes this a dangerous time for them. For this reason, many species gather in **colonies** on small islands and remote beaches, where they are fairly safe. Harp seals, however, do not need to return year after year to special islands or beaches. The floating sea ice provides an ideal place for giving birth.

Harp seals have four main breeding areas, including the White Sea off Russia and the Gulf of St Lawrence in Canada. In late winter, 12 months after **mating**, the females **haul out** onto the ice and give birth to a single offspring called a pup. The mothers feed their pups on their very rich, creamy milk for two weeks. Then they slip back into

A harp seal pup drinks its mother's milk. The pups are born with fluffy white fur to make them harder to see on the ice. People have killed many, many pups to make coats from their fur.

the sea, where they mate underwater with the males. The pups stay on the ice for another two weeks until they lose their infant coats of fluffy white fur. As the ice breaks up in the northern spring, they, too, take to the water.

Another seal that lives in the Arctic is the ringed seal. This small seal grows to only about 4½ feet (1.4 m) long and can be found throughout the region. Ringed seals usually breed on the thick ice found near the shores of many northern countries, including Norway, Iceland, and Canada. When the female is about to give birth, she usually hollows out a snow shelter. Here, she has her pup, which stays inside for the first month or two of its life.

Areas where harp and ringed seals can be found

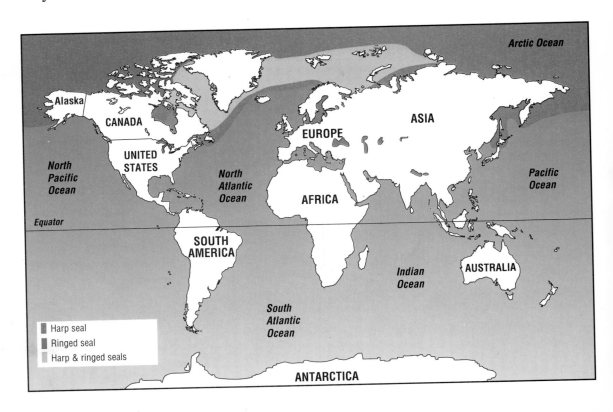

Harp seal
Ringed seal
Harp & ringed seals

Despite their efforts, seals are at risk when they are breeding on the ice. Arctic foxes and polar bears regularly sniff out ringed seal pups and kill them, and the Inuit have hunted seals for thousands of years. Far worse for the seals, though, has been commerical hunting. Until recently, hunters were killing thousands of white-coated harp seal pups each year for their fur. The protests of people from many countries brought an end to this slaughter, although older harp seals are still hunted.

Pollution is another serious threat to seals. Ringed seals in the Baltic Sea are harmed by chemicals called PCBs, which seriously affect their ability to breed. Commercial hunters are also planning to kill more harp seals each year in parts of their range. Only if pollution and hunting are strictly controlled will harp and ringed seals swim the northern seas of the future.

A ringed seal pup lies on the Arctic ice. Ringed seals are usually born in snow shelters, but sometimes mothers give birth in the open.

Area where the Mediterranean monk seal can be found

Monk Seals

As we have seen, harp and ringed seals live in some of the world's coldest waters, but long ago they probably preferred much warmer seas. Monk seals still live in warm-water **habitats**, chasing fish through the sunlit Mediterranean Sea and lazing on the **tropical** shores of Hawaii. Until recently there were monk seals in the Caribbean as well. However, as no Caribbean monk seals have been seen since 1952, they are probably now extinct.

The Mediterranean monk seal once lived throughout the Mediterranean area. It could be found from the Black Sea

A small octopus makes a tasty meal for a diving Mediterranean monk seal.

in the east to the Atlantic coast of Morocco in the west. The seal, which grows to about 8 feet 4 inches (2.6 m) long, could often be seen basking in the sun on rocks and beaches. The slightly smaller Hawaiian monk seal was equally common along the chain of islands stretching northwest from Hawaii to Midway Atoll.

People hunted monk seals, as they did all seals, and it was probably hunting that led to the Caribbean monk seal becoming extinct. In the case of the Mediterranean and Hawaiian monk seals, it is human disturbance for the most part that has put them at risk. Monk seals simply do not like to come into contact with people. In the Mediterranean,

Monk seals get their name from their dull gray or brown color, which reminded people of the color of the cloaks worn by monks. This is a Hawaiian monk seal.

Area where the Hawaiian monk seal can be found

the development of tourism has driven monk seals off their regular beaches. They now have to breed in sea caves on small islands. These are dangerous places, since during storms big waves often sweep the pups out to sea. Young pups cannot swim well in rough water, so they drown. The Mediterranean Sea is also heavily fished, and monk seals have to compete for food with commercial fishermen. For all these reasons, the Mediterranean monk seal is one of the most endangered mammals in the world.

The Hawaiian monk seal is not much better off. Like their Mediterranean relatives, Hawaiian monk seals are badly affected when humans move into their range. During World War II, the western Hawaiian islands were used by the military, and monk seal numbers fell. The large colony

A Hawaiian monk seal pup with a parent in Hawaii's National Wildlife Refuge. All monk seals like to raise their pups on quiet beaches, where they are safe from most of their enemies.

14

at Midway Atoll disappeared completely. In the 1950s, scientists found that more than one third of the pups born on one Hawaiian island were dying. When they are disturbed or frightened, monk seal mothers abandon their young, which then starve.

Monk seals are in serious danger of extinction. Today there are no more than 500 Mediterranean and only about 1000 Hawaiian monk seals left in the wild, but people are working hard to preserve them. The Hawaiian monk seals live in a special **sanctuary** that few people are allowed to enter. In the Mediterranean, **conservationists** are trying to protect their monk seals' breeding caves. If people can be persuaded to leave monk seals in peace, there is a chance that these shy creatures may survive.

Besides protecting the caves where Mediterranean monk seals breed, scientists want to teach fishermen and tourists about the seals, so they can take better care of them in the future.

Elephant Seals

Elephant seals are well named. Although a female elephant seal looks much like any other seal, the male is a huge creature. It can weigh more than 7000 pounds (3200 kg) and has a large nose that resembles an elephant's trunk.

There are two species of elephant seals: the northern and southern. The southern is the bigger, with males growing to almost 17 feet (5.2 m) long and females reaching about half this length. A male northern elephant seal, on the other hand, grows to only about 14 feet (4.3 m).

Elephant seals feed on deepwater fish, squid, and octopus, diving to 2000 feet (600 m) or more to catch them. A hunting elephant seal dives into deep water about

The northern elephant seal breeds on the beaches of California and Mexico. This bull's "trunk" acts a bit like a trumpet, enabling him to roar loudly to warn off rivals.

60 times a day. Each dive usually lasts about 20 minutes and is followed by a period of about three minutes on the surface. The seal needs this time to recover its breath. Scientists have learned all this by tracking elephant seals with radio transmitters. The longest dive ever recorded lasted two hours!

How does the seal manage to hold its breath for so long? The answer is that it doesn't. It breathes out before it dives and uses the oxygen stored in its blood and muscles to keep it going. All seals use this technique, but the elephant seals are among the best divers. People rarely see them at sea because they are nearly always underwater.

Areas where the northern elephant seal (brown) and southern elephant seal (green) can be found

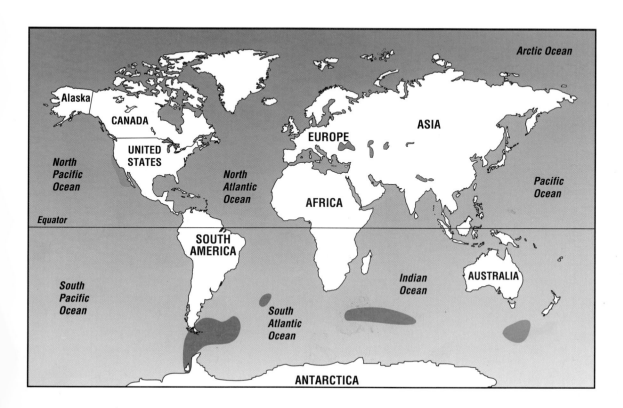

Elephant Seals

During the breeding season, female elephant seals gather in great colonies on breeding beaches, where they give birth to their pups. The same beaches have been used for many, many years. Male elephant seals also come ashore. Once the pups have been born, each male tries to mate with as many females as possible. The rivalry between males is intense and leads to bloody fights on the beaches. Only the biggest and most aggressive seals manage to mate, so they father nearly all the pups.

Like other seals, an elephant seal carries a lot of fat under its skin. In the nineteenth century, this fat was turned into lamp oil, and both species of elephant seals were heavily hunted. Luckily for the southern elephant seal,

Two young northern elephant seal bulls try out their strength on each other in the surf off California. One day, they may fight each other over females.

people could not reach some of the beaches where it came ashore. Many of this species survived the killing, and eventually its numbers started to rise again.

The northern elephant seal was an easier target. By 1890 there were only about 20 left, all living on Guadalupe Island off the Pacific coast of Mexico. By this time there was no longer much demand for lamp oil, and the seals managed to survive until they were given legal protection in 1922. Today about 125,000 northern elephant seals breed off the coasts of California and Mexico. But the northern elephant seal is not out of trouble yet. All these seals are descended from the 20 survivors on Guadalupe Island, so they belong to one big family. Breeding between different families is important to create a healthy population.

A group of southern elephant seals share their beach with a bunch of king penguins. Southern elephant seals breed on the islands that lie close to the icy continent of Antarctica.

Eared Seals

Eared seals differ from true seals in a number of ways. First, unlike true seals, they have visible ears. Also, eared seals push themselves through the water with their front flippers rather than their hind ones. And they have muscular, powerful shoulders and long necks. When on land, they can turn their hind flippers forward, so they act as feet. This makes them far more agile out of the water than true seals, and they can move with surprising speed. An angry eared seal often moves faster than a person, even over rough ground.

A Steller's sea lion – a species of eared seal. This male's thick mane will protect him in fights against rival males.

Another difference between true and eared seals is that eared seals do not have such a thick layer of fat. Instead, they have an undercoat of oily fur beneath their hairy coat. This keeps them warm on shore, and since it is waterproof it also helps keep the animals warm underwater. However, it does not work as well as a thick fat layer. If the animal dives deep, the water pressure squeezes the fur and makes it thinner. Eared seals rarely dive to great depths, though. They usually feed near the surface on fish, squid, and krill.

Eared seals are divided into two groups: the fur seals and the sea lions. Sea lions get their name from the thick manes of hair that males have on their necks and shoulders, which

Areas where eared seals can be found

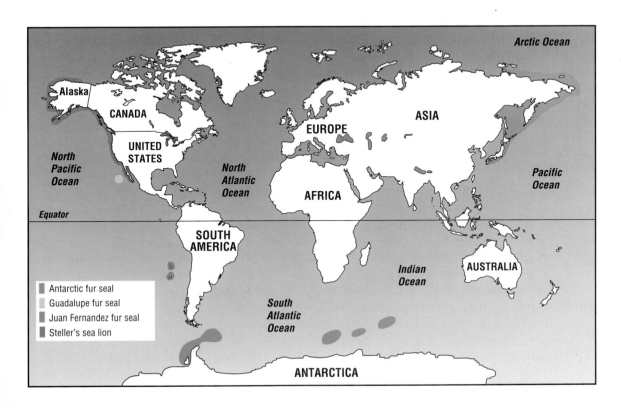

make them look a bit like lions. Male fur seals also have manes, but their **muzzles** are more pointed so they look less like big cats.

The biggest of all eared seals is Steller's sea lion, which is found on the coasts and islands of the North Pacific. A male can reach 10 feet (3 m) in length and weigh 2200 pounds (1000 kg). A female grows almost as long, but at about 600 pounds (272 kg) she is much slimmer. Most eared seals are a lot smaller than Steller's sea lion. A male Antarctic fur seal, for example, rarely weighs more than 440 pounds (200 kg). In the breeding season, male eared seals fight for **territories** on beaches. A powerful male with a big territory can attract a number of females, which form his **harem**. A weaker male may not breed at all.

A Galapagos sea lion has just given birth among the rocks. Baby seals are born easily and quickly.

Sea lion and fur seal breeding colonies can be huge. In the past this made it easy for hunters to kill these animals for their fur and oil. Many species were almost wiped out. People thought the Guadalupe fur seal of the North American Pacific had died out until it was spotted in 1928. Likewise, the Juan Fernandez fur seal of the South American Pacific was believed to be extinct for nearly 100 years, until it was discovered again in 1966.

Another fur seal that suffered badly from hunting was the Antarctic fur seal. There were hardly any of these fur seals left by 1908, when the last cargo of sealskins was shipped from the island of South Georgia in the South Atlantic. Since then Antarctic fur seals have staged a

A Juan Fernandez fur seal on Robinson Crusoe Island in the Pacific. Eared seals are fairly agile on land, which is why they have been trained to perform in circuses.

remarkable recovery. There may be more of them on South Georgia today than ever before. Sadly this is probably because the large krill-eating whales are now very scarce, so there is more krill in the sea for the seals to eat.

Meanwhile, other species of eared seals are still rare. Scientists are particularly worried about Steller's sea lion. This species was hunted heavily during the nineteenth century. Hunting then stopped and numbers increased. Since the 1960s, though, numbers have fallen again. There are many possible reasons why this has happened. Disease may be the cause, or a shortage of the fish the sea lions eat. Steller's sea lions also get tangled in fishermen's nets and drown. So far, scientists do not really know why Steller's sea lion is disappearing, but they are trying to find out.

Antarctic fur seal pups stay on their beach for about four months before going to sea for the first time. They must take care in the water, since leopard seals and killer whales hunt fur seal pups.

Walruses

The walrus is an extraordinary seal. It has the build of a heavyweight sea lion and a sea lion's forward-turning hind flippers, yet it swims more like a true seal.

There is only one species of walrus. Males can grow to well over 10 feet (3 m) long and weigh more than 2700 pounds (1225 kg). Females are almost as big. Walruses have extremely thick, folded skin over a thick layer of fat. This keeps them warm in the icy seas of their Arctic habitat. Both sexes have moustaches of sensitive bristles, but the most obvious features of any walrus are its tusks.

Walrus tusks, which are actually very large upper teeth, are made of a creamy white material called ivory. They are

Walruses hauled out off the coast of Alaska. These seals like to relax on rocky islands when there is no ice around for them to lie on.

usually about 22 inches (56 cm) long in males, while females' tusks are shorter and slimmer but still awesome. This is important, since the main reason walruses have tusks is to impress other walruses. Walruses are far more sociable than true seals. And like many social animals, they squabble over who is in charge. In order to decide who is the leader, walruses show off their tusks to one another. The biggest walrus with the longest tusks usually rules the others.

If showing off does not settle who is in charge, there may be a fight in which the rivals stab at each other with their tusks. Occasionally, walruses can be injured, but their

Areas where the walrus can be found

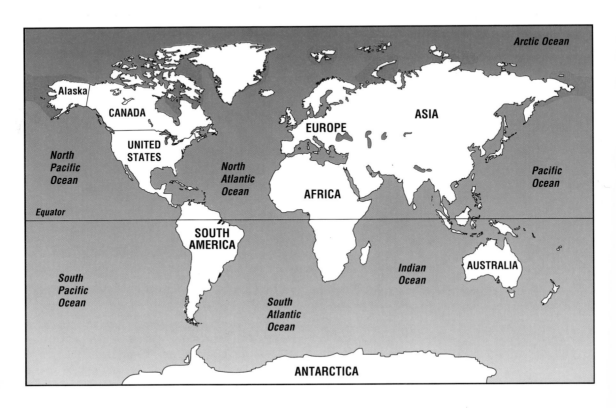

thick skin helps protect them. Also, the walrus that is losing usually retreats before any serious harm is done.

Walruses' tusks do have other uses. A walrus climbing out of the sea may haul itself onto the ice with its tusks. They also make excellent ice picks for smashing through new ice covering breathing holes. Some walruses use their tusks to kill smaller seals for food, but this is very unusual. Most feed only on clams and other shellfish, feeling for them in the dark on the seabed with their sensitive whiskers and digging them out. If a clam is deeply buried, a walrus may blast it out of its burrow with water squirted from its mouth at high pressure.

Walruses often come ashore to bask in the sun on rocks or ice. They gather in great herds and often lie on top of one another when there is not enough space. This habit

A walrus scratches at its nose with a flipper as it dozes in the sun. Like pigs, walruses use their noses to dig for food.

Walruses enjoy one another's company and often haul out in huge numbers, as here on Round Island in Alaska.

makes them an easy target for human hunters. The Inuit have hunted walruses for at least 10,000 years. They eat the meat, turn the skins into clothes, burn the oil, and use the ivory from the tusks for harpoons and tools and carvings

During the eighteenth century, though, large-scale killing of walruses began. Commercial hunters from Europe and North America almost wiped out a number of walrus colonies. Enough animals remained for the species to survive, though. Today the hunting of walruses is strictly controlled. As long as this is the case, the remarkable walrus should be safe.

Seals still face a number of threats in the modern world. Many are drowned in fishing nets, trapped in floating garbage, or frightened by human disturbance. Others face a food shortage because people overfish their waters. They die from pollution, too. On top of this, it is possible that commercial hunting of some seal species may increase. If seals are to survive, hunting and fishing need to be controlled. Human disturbance of shy species needs to stop, as does the pollution of the oceans. If we can do all this, we will preserve not only seals but many other marine animals and their world.

A baby harp seal shelters in a little cave in the ice. Harp seals can live to be 30 years old if they are given the chance.

Useful Addresses

For more information about seals and how you can help protect them, contact these organizations:

Center for Marine Conservation
1725 De Sales Street NW
Suite 500
Washington, D.C. 20036

The Cousteau Society
870 Greenbrier Circle
Suite 402
Chesapeake, Virginia 23320

Earthtrust
25 Kaneohe Bay Drive
Kailua, Hawaii 96734

Marine Mammal Stranding Center
P.O. Box 773
3625 Brigantine Boulevard
Brigantine, New Jersey 08203

U.S. Fish and Wildlife Service
Endangered Species and Habitat
Conservation
400 Arlington Square
18th and C Streets NW
Washington, D.C. 20240

World Wildlife Fund
1250 24th Street NW
Washington, D.C. 20037

Further Reading

Encyclopedia of Endangered Species Mary Emanoil (ed.) (Detroit: Gale Research, Inc., 1994)

Endangered Wildlife of the World (New York: Marshall Cavendish Corporation, 1993)

The Sea World Book of Seals and Sea Lions Phyllis Roberts Evans (San Diego: Harcourt, Brace, Jovanovich, 1986)

Seals The Cousteau Society (New York: Little Simon, 1993)

Seals Jinny Johnson (Columbus, OH: Highlights for Children, Inc., 1991)

Seals and Walruses Norman Barrett (New York: Franklin Watts, 1991)

Wildlife of the World (New York: Marshall Cavendish Corporation, 1994)

Glossary

Adapt: To change in order to survive in new conditions.

Colony: A large gathering of animals of the same species that often forms at breeding time.

Conservationist (Kon-ser-VAY-shun-ist): A person who protects and preserves the Earth's natural resources, such as animals, plants, and soil.

Extinct (Ex-TINKT): No longer living anywhere in the world.

Habitat: The place where an animal lives. For example, the walrus's habitat is Arctic waters.

Harem (HAIR-uhm): In some species of seals, a male gathers a group of females that he mates with and guards. This group is known as his harem.

Haul out: When a seal drags itself out of the water and onto land.

Mammal: A kind of animal that is warm-blooded and has a backbone. Most are covered with fur or have hair. Females have glands that produce milk to feed their young.

Mate: When a male and female get together to produce young.

Muzzle: The protruding part of the face made up of the nose and jaw. Animals with muzzles include seals, bears, and dogs.

Pollution (Puh-LOO-shun): Materials, such as garbage, fumes, and chemicals, that damage the environment.

Prey: An animal that is hunted and eaten by another animal.

Range: The area in the world in which a particular species of animal can be found.

Sanctuary (SANK-chu-wer-ee): A safe place.

Species: A kind of animal or plant. For example, the harp seal is a species of seal.

Territory: An area that an animal will defend against others of its own kind.

Tropical: Having to do with or found in the tropics, the warm region of the Earth near the Equator.

Index